The Secret Language of Dogs

Sarah Hodgson

PRENTICE HALL

Printed in the United States of America

10 9 8 7 6 5 4 3

Adapted from: *The Complete Idiot's Guide to Choosing, Training and Raising a Dog* by Sarah Hodgson

ISBN 0-13-020731-4

PRENTICE HALL
Paramus, NJ 07652

On the World Wide Web at http://www.phdirect.com

Not English, Doglish!

In This Section

- English to Doglish translations
- The irony of negative attention
- Positive attention rules
- How to be a Doglish interpreter

There is more to your little fur ball than blinking eyes and four paws! A fascinating creature in his own right, he needs you to understand where he's coming from before he'll give you his full cooperation. No matter how many stars you wish on, he'll never understand what it is to be human. Before you learn specific training techniques, you need to explore your dog's individual personality and learn his language.

Until you can think *with* and *not against* your dog, you can't really train him properly. It's impossible for your dog to be human, no matter how much you work together. So how do you think like a dog? It's quite an adventure.

The Leadership Principle

Dogs have a lot of team spirit. This is often referred to this as their "pack" instinct, but I like to think of it in "team" terms. Team consciousness, and the canine psyche have a lot in common. Teams focus on winning; each player works for it, wants it, thinks about it, and strives for it. There are also some other, less obvious factors that determine a team's success. Three come to mind immediately: cooperation, structure, and mutual respect. Without these, even a group of phenomenal players would be pure chaos. A good team is organized so everyone knows who's in charge and what's expected from them. And should someone get in trouble or become hurt, he can trust that another teammate will help out.

Dogs live their entire lives, their every waking moment, by team structure. Instead of winning, however, their mantra is survival. And to personalize it one step further, you and your family are their team. For your dog to feel secure and safe, he *must* know who's in charge. It's your job to teach him what you expect from him in your home. Do you have more than one person in your household? In dogland, teams are organized in a hierarchy, so you must teach your four-legged friend that two-legged dogs are the ones in charge. This will take some cooperation on everyone's part, but it's very do-able.

> If you don't organize the team hierarchy, your dog will, and that can be a real nightmare. If your dog has the personality to lead, you'll be living in a very expensive doghouse under dog rule. If your dog doesn't have what it takes but feels he must lead because no one has applied for the job, your house will be one big headache; dogs in this state are very hyper and confused.

So how do you organize your team and teach your dog the rules? The first step is understanding what motivates your dog's behavior. Then you need to master his communication skills. It might sound like hard work, but it's quite fascinating. Your dog will respond to you more willingly if you make the effort to understand and learn his language. With an ounce of effort, a little time, and some structure, you'll earn your dog's respect, cooperation, and trust. Plus, you'll have a teammate who'll be at your side when the cards are down. You can't beat that bargain!

The Attention Factor

Dogs are motivated by attention. They live for it, love it, and will do anything to keep the spotlight focused on them. Does this remind you of a three year old? Well, add to this similarity the fact that they don't care whether the attention is negative or positive.

The Irony of Negative Attention

Picture a very excited jumping dog. You're trying to read the paper calmly, but he wants your attention. What if you tried to correct the dog by pushing him down and screaming "Off!"? In all likelihood, the dog will jump again. Do you know why? Because you just gave him attention. Attention in a dog's mind includes anything from dramatic body contact to a simple glance. Yes, even looking at your dog will reinforce his behavior.

Does this blow your mind? Though it may sound far-fetched at first, it's actually pretty elementary. Dogs think of us as other dogs. If they get excited and then we get excited, we're following their lead. The fact that you might be upset with their behavior just doesn't register. Being upset is a human emotion. Excitement and body contact is a dog thing. Even if you push your dog so hard that he stops and slinks away, the only thing you've accomplished is scaring your dog. And who wants to train a dog through fear? Trust me, there's a better way.

Let me give you another example. What happens if a dog grabs a sock and everyone in the household stops to chase him? Dog party? You bet. Because the puppy views everybody as a dog, he's thinking, "What fun!" as he's diving behind the couch and under the table. Chasing doesn't come across as discipline; it comes across as prize envy—"Whatever I have must be really good because everyone wants it!"

Now you're wondering how to resolve these problems. We'll get there, I promise, but for now keep in mind that out-of-control negative attention reinforces the very behavior you're trying to change. I know it's frustrating. But read on and be patient.

Power of Positive Attention

When I ask my clients what they do when they catch their dog resting or chewing on a bone quietly, most say, "Nothing. It's a moment of peace." I appreciate such honesty; however, that's when they ought to be showering their dog with attention. Not wild, twist-and-shout, hoot-and-holler attention, just calm, soothing, loving attention that makes them smile inside. A soft whispering praise is best mixed with a massage-like pat. Remember my mantra?

You dog will repeat whatever you pay attention to.

So you decide. What would you rather have? A dog that stays by your side with a chew bone or a frantic sock stealer that races around the house like a maniac? If you like the sock stealer, close the book. But if the bone-chewer image appeals to you, stick with me—we're going places!

Doglish

One more thing before we jump into the how-to's of training: I have to teach you the dog's language. To be the best teacher, you need to be fluent in *Doglish*. Give your family or friends a lesson too and encourage consistency.

Doglish consists of three elements:

- Eye contact
- Body language
- Tone

Words, feelings, and lengthy explanations don't count anymore. Complex reasoning is impossible for your dog to follow. Dogs are so innocent in their simplicity, it's beautiful.

Eye Contact

If you're constantly looking to your dog in stressful situations (someone's at the door or the dog's stealing the dish rag) and are having trouble encouraging your dog to pay attention to you, guess what? Your dog thinks you are depending on him to be the leader. He thinks you want him to make all the judgment calls. I know you're wondering how to handle these situations. In just a few moments, I'll tell you. But for now, understand that to train your dog, you must encourage him to look to you for direction. To leave you with an off-shoot of my attention statement:

You reinforce whatever you look at.

Look at a well-behaved dog and guess what you'll have?

Body Language

Body language is a funny thing. Imagine this...your dog becomes excited and hyper when company arrives at the front door. Desperate to save face, you start shouting and pushing your dog as the company is fending the two of you off with their coats. You try every possible command—"Sit Boomer! Down! Off! Bad dog!"—but to no avail. The whole arrival scene is one big fiasco.

What's happening dear readers? Who has copied whom? Whose body language has mimicked the other's? Are you feeling silly yet? Body language is an integral part of Doglish. Play, tension, relaxation, they all have different postures. Going on the knowledge that your dog thinks you're a dog and doesn't quite grasp the *"I'm pushing you frantically because I'm unhappy with your greeting manners"* concept, what do you think you're communicating to your dog? You're the one who copied his body language. As you blaze the training trail, remember these three things:

- Stand upright and relax when directing your dog. I call this the *peacock position*.
- Don't face off or chase your dog when you're mad. To your dog, you'll look like you're playing.
- When trying to quiet or direct your dog, stand in front of him and stay calm.

Always remember, you set the example. I hear two questions ringing throughout the pages already:

How on earth can this be done?

Can't I ever get down and play or cuddle with my dog?

We are getting there, and certainly you can get down and cuddle or play; that's one of the biggest perks in having a dog. But don't do it when your dog's in a mischievous mood or you'll be asking for trouble.

Imagine a peacock—beautiful and proud, chest out, confident, and in control. When giving your dog direction or a command, throw your shoulders back and stand tall just like a peacock. Tell your family and friends about this **peacock position** and start strutting your stuff!

Tone

If your dog thinks of you as another dog and you start yelling, what does he hear? Yes, he hears barking. And would barking calm excitement or increase it? Right again, it would increase it. Now some of you may have a dog that backs off from a situation when you yell, though he'll probably repeat the same behavior later. That's because yelling frightened him. He backed off because he was afraid of you, not because he understood. Yelling at any angle is just no good. So what works? Well, we'll get there in just a few moments, but before we begin, there are three tones you should commit to memory. I call them the three Ds:

- **Delighted Tone**. Your praise tone should soothe your dog, not excite him. Find a tone that makes him feel warm and proud inside.
- **Directive Tone**. Use this tone for your commands. It should be clear and authoritative, not harsh or sweet. Give your commands once from the peacock position.
- **Discipline Tone**. I'm not much of a disciplinarian. My approach encourages more structure than strictness, but you should have a few tones that tell your dog to back off or move on. I use "Ep-Ep" a lot. I'll go into more detail later, but the word doesn't matter as much as the tone. The tone should be shameful or disapproving, like "How could you" or "You better not touch that..." Discipline has more to do with timing and tone than your dog's transgressions.

> If you bend over when giving your dog a command, don't be surprised if he doesn't listen. You're doing the doggy equivalent of a **play bow** (a posture that invites a game). To put it in human terms, if you asked me to have a seat while you were hunched over looking at the floor, I'd think less of where to sit and wonder what on earth you were looking at. Stand tall and proud like a peacock when giving your dog directions.

> Do you have kids? Since they'll copy you, overenunciate all of your commands so they'll learn to pronounce them properly. For example, instead of saying, "Sit," say "SIIIT." When your kids copy your intonations, the control will transfer from you to them!

As you can see, Doglish is quite different from English. Many people assume that their dog understands them when, in fact, he's often picking up the opposite message. If you chase your table snatcher, your English is saying, "How dare you," but your Doglish is saying, "PARTY!" Remember, every interaction you have with your dog gets translated into Doglish. From now on, you're being watched from a canine's eyes, so you had better start acting like one!

> Don't repeat your commands. Dogs don't understand words as words; they learn to respond to specific tones and syllables. If you say "Sit, sit, sit, Boomer sit!" that's what Boomer will learn. If you want your dog to listen when you give the first command, make sure you give it only once and reinforce your expectations by positioning your dog.

The Least You Need To Know

- Dogs view humans as other dogs. They relate to you, your family, and close friends as their team.
- Each member of the team is ranked according to leadership and organizational ability. Two-legged dogs should lead four-legged ones!
- Dogs don't care whether attention is negative or positive. If an action gets a reaction, a dog will repeat it. If you pay attention to a calm, well-mannered creature, then that's what you'll get. Unfortunately, the opposite is also true.
- Dogs don't understand English. They have their own unique language, called Doglish. To be a good teacher, you must learn their language and speak it as you train them.

Other Horrors

In this Section

- How to curb nipping, chewing, and barking
- When your dog's idea of fun isn't fun for you
- Four paws on the floor: redirecting your jumper!

Order in the house! Order in the house! Sometimes you'd like to yell it and have it be so. Teaching household etiquette may be the most trying time for you and your dog. While your dog is trying to figure out what you want, you've been reduced to pleading for cooperation: "I'll give you a biscuit. Biscie Boomer. Please, Boomer, please? I'm late for work."

It's embarrassing. You need some guidelines to get on the right track. In this section, I'll cover chewing, the infamous grab-n-go, jumping, and nipping.

Stop Chewing!

Chewing is a dog thing. It's nothing personal. They don't know a stick from furniture or a doll's head from a chestnut. Fortunately, they can be rehabilitated. If you have a habitual chewer on your hands, however, you'll need to be patient and use some of the tried-and-true techniques described next.

Just like kids, pups are curious about the world around them and they love to explore. Kids use their hands; puppies use their mouths. Additionally, pups between 3 1/2–8 months are teething. Your puppy might chew on the furniture or your favorite shoes to alleviate discomfort. To ward off possible destruction, supply and encourage the use of appropriate chew toys.

Buy Stock in Bitter Apple®

Bitter Apple® is nasty tasting stuff you can buy at most pet stores that you can spray on items to prevent your dog from chewing them. If you notice your dog chewing on the furniture surrounding her station, spray everything but her bed and bone. Believe it or not, some dogs like Bitter Apple®. If this is your dog, try a Tabasco sauce mixture.

Provide One Main Toy

Having too many objects to choose from can confuse your dog. Pick a bone or toy that will satisfy your dog's penchant for chewing, buy multiples of that item, and spread them around the house for quick access. Do the same for play toys.

Avoid Prize Envy Confrontations

Don't yell at your dog after she's begun or after she's finished chewing. It's too late. If you chase a dog who has something in her mouth, she'll be thinking, "Wow, what a great prize...everybody wants to take it from me!" (This is called prize envy.) Instead, learn about the treat cup and use if effectively.

Create Treat Cups

Making a *treat cup* is easy. Break up your dog's favorite treats in a cup. Shake the cup and offer your dog a treat. Continue this until your dog associates the sound of the cup with getting a treat. Now spread treat cups all over your home and be consistent. Use the same kind of cups. Party cups or deli containers work best.

Anytime your dog is chewing on an *acceptable* object, go over with the treat cup, say "Out," offer her a treat, and leave. When you dog's eating a meal, shake a cup, say "Out," offer her a treat, and leave. Now that you've communicated that your approach is not threatening, the next time your dog grabs something, find a treat cup and say "Out." Treat all objects she grabs, good or bad, as treasures and she'll be much more cooperative. Praise her when she releases the object and help her find a chew toy: "Where's your bone?"

Jaws Junior: Nipping and Mouthing

Mouthing and nipping are two different issues. *Nipping* is a puppy thing; it's interactive and playful. *Mouthing* is a lesser infraction; it's more of a communication skill to get you to do a particular thing. Less pressure, less annoying, but still not a charming habit. Nipping dogs are bossy and manipulative and need a firmer regime.

Mouthing is an attention getting behavior. If your dog uses it to communicate a need to go out, respond. If your dog is mouthing you for a pat, please ignore it. Pretend she isn't there. If she becomes too annoying, get Binaca Mouth Spray® and spritz her discreetly, hiding the Binaca in your hand, avoiding all eye contact, comments, or pushing.

Nipping with sharp little needle teeth can hurt! It's another one of those dog-things that you'll need to refocus. Consider this: When your puppy still hung out with her littermates, she nipped during play and to determine her rank. She also soft-mouthed her mother affectionately. When you bring your puppy home, what happens? This behavior continues. What your puppy wants to know is who's puppy and who's not. This determines the type of mouthing or nipping: soft or playful. Usually, everyone gets categorized as a puppy. *Why?* Well for starters, most people pull their hand away when nipped. To a human, it's self-defense. To a pup, it's an invitation to play. Even if you were to correct your young puppy, she wouldn't get it; it's like correcting a one-year-old baby for pulling your hair. So

Dogs interpret discipline as confrontational play. Excessive physical corrections will result in aggression. Be wise; stay cool.

what should you do? Good question. Your approach will depend on your puppy's age.

Younger Than 14 Weeks

Young puppies mouth a lot. They mouth when playing; they also mouth to communicate their needs, just like a baby cries. If your puppy starts mouthing, ask yourself: Is she hungry or thirsty? Does she need to eliminate? Is she sleepy? Does she need to play? Next, follow the checklist to control mouthing and nipping!

- Whenever your puppy licks you, say "kisses" and praise her warmly.
- Hold your attention when your puppy nips softly. Keep your hand still. Don't forget, hand withdrawal is an invitation to play and nip harder.
- If your puppy starts biting down, turn on her quickly, say "Ep, Ep!" and glare into her eyes for two seconds. Go back to your normal routine.
- Remember, puppies nip when they feel needy (just like a baby cries). If your puppy won't let up, ask yourself if she wants something, like an outing, exercise, or a drink. If all checks out and your puppy won't quit, crate or isolate her with a favorite bone. Do not scold your puppy as you isolate her. Calmly place your puppy in her area.

Training Your Puppy around Kids

Kids act a lot like puppies. They're always on the floor and into everything. If you have children, teach your puppy not to mouth them from the start. Here's how.

Leave your puppy on a one-foot long nylon leash whenever she's with your children. If she starts playing too rough, pick up the leash, snap back, and say "Ep, Ep." If you're still having trouble, buy a super soaker or plant mister and fill it with water and vinegar. Spray your dog discreetly when she starts getting riled up.

If all else fails, give the puppy a time-out attached to you, stationed, or crated. Help the kids see that their restlessness leads to puppy-withdrawal.

Older Puppies

Do you have a Peter-Pan pup, one who still nips past her time? Well, the buck stops here. After 14 weeks, there's no excuse. If your puppy's still nipping, it's likely that you've taught her, so it's you who'll have to mend your ways:

- Stop all challenge games. These games include wrestling, tug-of-war, chasing your dog around, and teasing. You're sending the wrong message.
- Correct all nipping, whether it's a bite on your arm or a nibble on your finger. Teeth do not belong on human skin, period.
- Purchase a few weapons to use in defense, such as Binaca Mouth Spray®, Bitter Apple® spray, or a long-distance squirt gun.
- Give yourself something to grab. If your dog's not wearing the Teaching Lead®, place a short lead onto her buckle collar.

> No tug-of-war or rough wrestling matches please! These games teach dogs to clamp down hard on any object and challenge. This could be a leash, the laundry, your shirt, or even your skin.

> If things get out of hand, avoid yelling at the kids. It sounds like barking to your dog and ups the fun ante. Calmly station, crate, or isolate your puppy until she mellows

- If your dog begins to mouth, turn to her, use a lead or collar to snap her head from your body, and say "Ep, Ep!" Glare at her for a second and then go back to business as usual.
- If she continues to nip, ask yourself: Do I look convincing? Am I snapping or pulling? (Pulling encourages play.) Is my dog taking me seriously? You may need more training before you earn her respect.
- Carry Binaca Mouth Spray®. Spritz your dog whenever she's mouthing and say "Ep, Ep!" Spray her nose once or twice. After that, you can spray the air above her head; the blasting sound will warn her off.

Curing the Chase and Nip

There are two categories to cover here:

- **The bathrobe assault.** If your dog's a clothing grabber, dilute some Bitter Apple® spray in a plant mister and carry it with you when you suspect your dog will pull this assault. Do not turn and face your dog when she jumps; this is interpreted as a challenge. Without looking or responding, spray your dog and continue walking. If this problem persists, get help now. It can develop into post-puberty aggression. No joke.
- **The child chaser.** Kids running around the yard, apartment, or house are a big temptation. If you were a dog, you'd be jumping and nipping too. Because you can't teach kids to stop being kids, you need to help your dog control her impulses. Put your dog on the Teaching Lead® and ask the kids to race around in front of you. Anytime your dog looks tempted to lunge, snap back and say "Shhh." Repeat as often as necessary to gain control.

Two Feet, the Joyous Jumper

Everybody knows a jumper—a knock-you-over-when-you-come-in jumper, a muddy-paws-on-the-couch jumper, a counter cruiser (a dog who likes to sniff along counter tops). Jumping is a sure-fire attention getter. So what gives? The first step in solving your problem will be to understand how it became a problem in the first place. Once again, your dog's not to blame. Let's hop into his paw and see what's going on.

Dogs see us as other dogs. Eye contact is a big method for canine communication. Our eyes are up, so to be gracious and greet us properly, dogs must jump. The first time this happens, often in puppyhood, a hug follows. "Isn't that cute?" After about the tenth jump, it's not so cute. So the dog usually gets a shove. But what's a shove to a dog? Confrontational play. The dog jumps higher and harder the next time. So the human tries a little toe stepping, paw grabbing, yelling, all with the same effect; dogs think jumping is very interactive and very fun.

Counter jumping is another favorite pastime. After all, we're looking at the counter constantly, so why shouldn't the dog do so as well? When a dog jumps up, the human reacts by shouting and shoving. The dog interpretation? Prize envy. The dog thinks, "Whatever I was reaching for must be excellent because everybody raced over for it." So the dog reconsiders. He jumps when your back is turned or you're out of the room. Is this behavior spiteful? No, just plain smart. Now let's dissect and correct this problem one jumping situation at a time.

Your Homecoming

The best way to remedy jumping when you come home is to ignore your dog. Try it for a week. Come home and ignore your dog until she's given up the jumping vigil. Keep a basket of balls or squeaky toys by the door. When you come in, toss one on the ground to refocus your dog's energy. If your dog's crated, don't let her out immediately; wait until she's calm.

Dogs mimic their leaders' energy levels. If you come home to an excited dog and you get excited, what message are you sending? Instead, come in calm and wait to greet your dog until she's settled down too!

If you have a big dog or a super persistent jumper, put on an overcoat to protect yourself. Whether it takes two minutes or 20, go about your business until your dog calms down.

Do you have kids? Tell them to look for rain and do the same. Cross your arms in front of your chest and look to the sky. Don't look down until the coast is clear. Consistency is key. If one family member follows the program but the others encourage jumping, your dog will jump-test all visitors.